Math Mammoth

Grade 1 Tests & Cumulative Reviews

for the complete curriculum
(Light Blue Series)

Includes consumable student copies of:

- Chapter Tests
- End-of-year Test
- Cumulative Reviews

By Maria Miller

Contents

Grade 1, Chapter 1

End-of-Chapter Test

Instructions to the student:

Answer each question in the space provided.

Instructions to the teacher:

My suggestion for grading the chapter 1 test is below. The total is 28 points. Divide the student's score by the total of 28 to get a decimal number, and change that decimal to percent to get the student's percentage score.

Question #	Max. points	Student score
1	6 points	
2	8 points	
3	6 points	

Question #	Max. points	Student score
4	4 points	
5	4 points	
Total	28 points	

Chapter 1 Test

1. Add.

a.
$$\begin{array}{r} 2 \\ +\ 4 \\ \hline \end{array}$$

b.
$$\begin{array}{r} 8 \\ +\ 1 \\ \hline \end{array}$$

c.
$$\begin{array}{r} 6 \\ +\ 2 \\ \hline \end{array}$$

d.
$$\begin{array}{r} 8 \\ +\ 2 \\ \hline \end{array}$$

e.
$$\begin{array}{r} 5 \\ +\ 4 \\ \hline \end{array}$$

f.
$$\begin{array}{r} 6 \\ +\ 3 \\ \hline \end{array}$$

2. Compare. Write $<$, $>$, or $=$.

a. 5 ☐ 6	c. $6+1$ ☐ 10	e. 3 ☐ $1+0$	g. 8 ☐ $6+3$
b. 2 ☐ 9	d. $2+3$ ☐ 5	f. 10 ☐ $0+6$	h. 7 ☐ $2+5$

3. Draw more. Write an addition sentence.

a. _____ + _____ = 10

b. _____ + _____ = 7

c. _____ + _____ = 10

4. Find the missing numbers.

a. $2 +$ _____ $= 7$ b. $1 +$ _____ $= 4$ c. $4 +$ _____ $= 10$ d. _____ $+ 7 = 9$

5. Solve the word problems.

a. Anna has seven stuffed animals and Abby has three. How many do they have in total?

b. Andy has eight pairs of shorts. Two of them are in the wash. How many are not?

Grade 1, Chapter 2

End-of-Chapter Test

Instructions to the student:

Answer each question in the space provided.

Instructions to the teacher:

My suggestion for grading the chapter 2 test is below. The total is 30 points. Divide the student's score by the total of 30 to get a decimal number, and change that decimal to percent to get the student's percentage score.

Question #	Max. points	Student score
1	4 points	
2	4 points	
3	6 points	

Question #	Max. points	Student score
4	16 points	
Total	30 points	

Chapter 2 Test

1. Write the fact family to match the picture.

_____ + _____ = _____   _____ + _____ = _____

_____ − _____ = _____   _____ − _____ = _____

| 8 |
| :●● / :●●: |

2. **a.** Write a subtraction sentence that matches the
   addition $5 + 4 = 9$, using the same numbers.   _____ − _____ = _____

   b. Solve the addition $6 + $ _____ $= 10$ and
   write a subtraction that matches the addition.   _____ − _____ = _____

3. **a.** There are 9 animals playing in the yard. Three are
   dogs and the rest are cats. How many cats are there?

   b. Lisa has four more balls than Kelly.
   Kelly has five balls.
   Draw Kelly's and Lisa's balls.

   c. Five robins and two sparrows are feeding on seeds.
   Two more robins fly in. Now how many more
   robins are there than sparrows?

4. Find the missing numbers.

a. $4 + $ _____ $= 6$	**b.** $9 − $ _____ $= 3$	**c.** $9 − 0 = $ _____	**d.** $6 − 5 = $ _____
$1 + $ _____ $= 8$	$7 − $ _____ $= 5$	$7 − 1 = $ _____	$3 − 2 = $ _____
$5 + $ _____ $= 10$	_____ $− 1 = 6$	$9 − 7 = $ _____	$4 − 4 = $ _____
$6 + $ _____ $= 9$	_____ $− 2 = 4$	$8 − 2 = $ _____	$10 − 7 = $ _____

Grade 1, Chapter 3

End-of-Chapter Test

Instructions to the student:

Answer each question in the space provided.

Instructions to the teacher:

My suggestion for grading the chapter 3 test is below. The total is 27 points. Divide the student's score by the total of 27 to get a decimal number, and change that decimal to percent to get the student's percentage score.

Question #	Max. points	Student score
1	4 points	
2	5 points	
3	5 points	
4	4 points	

Question #	Max. points	Student score
5	6 points	
6	3 points	
Total	27 points	

Chapter 3 Test

1. Name the numbers (with words).

   a. 1 ten 6 ones _____

   c. 7 tens 8 ones _____

   b. 5 tens 1 ones _____

   d. 9 tens 0 ones _____

2. Fill in the missing numbers on the number line.

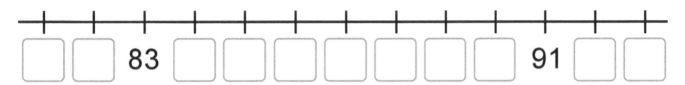

   ☐ ☐ **83** ☐ ☐ ☐ ☐ ☐ ☐ **91** ☐ ☐

3. Break the numbers into tens and ones.

a. $45 = \quad 40 \quad + \quad 5$	**b.** $52 = \underline{\quad\quad} + \underline{\quad}$	**c.** $97 = \underline{\quad\quad} + \underline{\quad}$
$86 = \underline{\quad\quad} + \underline{\quad}$	$32 = \underline{\quad\quad} + \underline{\quad}$	$19 = \underline{\quad\quad} + \underline{\quad}$

4. Do the same the other way around! Add.

a. $20 + 9 = \underline{\quad\quad}$	**b.** $5 + 70 = \underline{\quad\quad}$	**c.** $2 + 80 = \underline{\quad\quad}$	**d.** $1 + 90 = \underline{\quad\quad}$

5. Put the numbers in order from smallest to greatest.

a. 75, 71, 57	**b.** 69, 98, 96	**c.** 81, 84, 49
$\underline{\quad} < \underline{\quad} < \underline{\quad}$	$\underline{\quad} < \underline{\quad} < \underline{\quad}$	$\underline{\quad} < \underline{\quad} < \underline{\quad}$

6. Compare and write $<$, $>$, or $=$.

   a. $65 \ \boxed{} \ 5 + 60$ **b.** $43 \ \boxed{} \ 60 + 4$ **c.** $90 + 3 \ \boxed{} \ 30 + 9$

Grade 1, Chapter 4

End-of-Chapter Test

Instructions to the student:

Answer each question in the space provided.

Instructions to the teacher:

My suggestion for grading the chapter 4 test is below. The total is 32 points. Divide the student's score by the total of 32 to get a decimal number, and change that decimal to percent to get the student's percentage score.

Question #	Max. points	Student score
1	14 points	
2	6 points	

Question #	Max. points	Student score
3	12 points	
Total	32 points	

Chapter 4 Test

1. Find the missing numbers.	a. 2 + _____ = 8	b. 2 + _____ = 9	c. 2 + _____ = 10
	3 + _____ = 8	3 + _____ = 9	3 + _____ = 10
	1 + _____ = 8	5 + _____ = 9	6 + _____ = 10
	4 + _____ = 8	1 + _____ = 9	1 + _____ = 10
d. $8 - 5 =$ _____	**e.** $9 - 8 =$ _____	**f.** $10 - 4 =$ _____	**g.** $7 - 3 =$ _____
$8 - 7 =$ _____	$9 - 9 =$ _____	$10 - 9 =$ _____	$6 - 5 =$ _____
$8 - 2 =$ _____	$9 - 7 =$ _____	$10 - 7 =$ _____	$7 - 6 =$ _____
$8 - 3 =$ _____	$9 - 4 =$ _____	$10 - 8 =$ _____	$6 - 3 =$ _____

2. **a.** Sally has seven coins. Liz has three coins. Today, Liz found five more coins.
Now who has more coins?
How many more?

   b. Dan had two boxes of nails. Then he bought four more boxes of nails.
The next day he gave three boxes to the neighbor.
How many boxes of nails does Dan have now?

3. **a.** Complete. Then connect with a line the facts from the same fact family.

   b. Complete. Then connect with a line the facts from the same fact family.

_____ $- 4 = 3$	$8 - 3 =$ _____
$3 +$ _____ $= 5$	_____ $+ 3 = 7$
$8 -$ _____ $= 5$	$5 - 2 =$ _____

$2 +$ _____ $= 6$	_____ $- 4 = 3$
$7 -$ _____ $= 4$	_____ $- 6 = 3$
_____ $+ 3 = 9$	$2 + 4 =$ _____

Grade 1, Chapter 5

End-of-Chapter Test

Instructions to the student:

Answer each question in the space provided.

Instructions to the teacher:

My suggestion for grading the chapter 5 test is below. The total is 34 points. Divide the student's score by the total of 34 to get a decimal number, and change that decimal to percent to get the student's percentage score.

Question #	Max. points	Student score
1	8 points	
2	12 points	
3	10 points	

Question #	Max. points	Student score
4	4 points	
Total	34 points	

Chapter 5 Test

1. Write the time using the expressions *o'clock* and *half past*.

a. _____

_____

b. _____

_____

c. _____

_____

d. _____

_____

2. Write the time in two ways: using *o'clock* or *half past*, and with numbers.

a. _____

_____

_____ : _____

b. _____

_____

_____ : _____

c. _____

_____

_____ : _____

d. _____

_____

_____ : _____

3. Write the time for a half-hour and an hour later from the given time. Use numbers.

Now it is:	a. 6:00	b. 9:30	c. 10:00	d. 4:30	e. 12:30
a half-hour later, it is:					
an hour later, it is:					

4. Fill in either AM or PM.

a. Anna wakes up. It is 7 _____.	b. Anna plays after lunch. It is 3 _____.
c. Anna sleeps. It is dark. It is 3 _____.	d. Have an evening snack! It is 7 _____.

Grade 1, Chapter 6

End-of-Chapter Test

Instructions to the student:

Answer each question in the space provided.

Instructions to the teacher:

My suggestion for grading the chapter 6 test is below. The total is 9 points. Divide the student's score by the total of 9 to get a decimal number, and change that decimal to percent to get the student's percentage score.

Question #	Max. points	Student score
1	2 points	
2	2 points	
3	3 points	

Question #	Max. points	Student score
4	2 points	
Total	9 points	

Chapter 6 Test

1. The two shapes are put together.
   What new shape is formed?

   a. _____

   b. _____

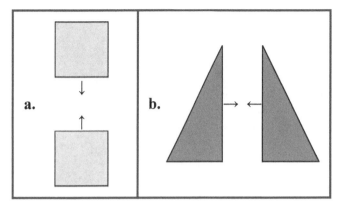

2. Color the triangle red, one rectangle green and the other purple, the squares yellow, and the circles light blue.

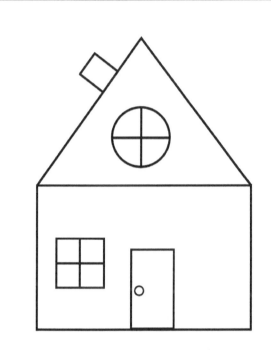

3. Join these dots <u>carefully</u> with lines. Use a ruler. What shape do you get?

   _____

   Measure the sides of your shape in inches.

4. Draw a line that is:

   a. 4 inches

   b. 12 centimeters

Grade 1, Chapter 7

End-of-Chapter Test

Instructions to the student:

Answer each question in the space provided.

Instructions to the teacher:

My suggestion for grading the chapter 7 test is below. The total is 25 points. Divide the student's score by the total of 25 to get a decimal number, and change that decimal to percent to get the student's percentage score.

Question #	Max. points	Student score
1	3 points	
2	6 points	
3	2 points	

Question #	Max. points	Student score
4	8 points	
5	6 points	
Total	25 points	

Chapter 7 Test

1. Add and subtract.

a. 22 + 4 = _____	**b.** 40 + 30 = _____	**c.** 90 − 20 = _____
41 + 5 = _____	76 + 10 = _____	80 − 70 = _____

2. Add. First, make a new ten with some of the little dots.

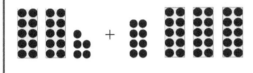 

a. 25 + 38 = _____	**b.** 14 + 25 = _____	**c.** 27 + 27 = _____

3. Add. You can use the trick with nine and the trick with eight.

a. 9 + 9 = _____	**b.** 4 + 9 = _____	**c.** 8 + 5 = _____	**d.** 8 + 7 = _____

4. Add and subtract. Write the numbers under each other.

    a. 20 + 57   **b.** 78 − 44   **c.** 45 + 13   **d.** 87 − 20

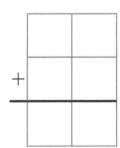 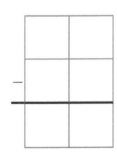 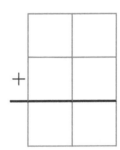 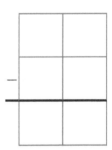

5. Jake had 6 dollars and Jim had 12. Then, Jake got 10 dollars more.

Now who has more money?

How many dollars more?

Grade 1, Chapter 8

End-of-Chapter Test

Instructions to the student:

Answer each question in the space provided.

Instructions to the teacher:

My suggestion for grading the chapter 8 test is below. The total is 13 points. Divide the student's score by the total of 13 to get a decimal number, and change that decimal to percent to get the student's percentage score.

Question #	Max. points	Student score
1	6 points	
2	3 points	

Question #	Max. points	Student score
3	4 points	
Total	13 points	

Chapter 8 Test US Money

1. How much money? Write the amount in cents.

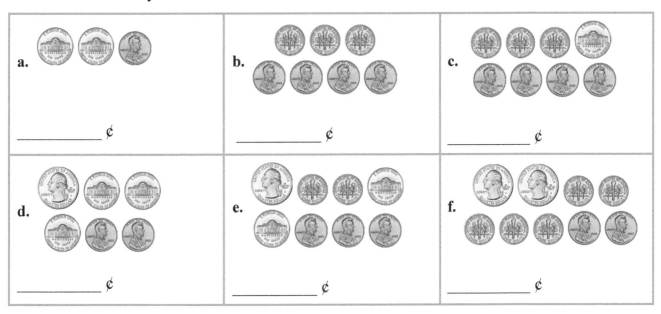

a. _____ ¢

b. _____ ¢

c. _____ ¢

d. _____ ¢

e. _____ ¢

f. _____ ¢

2. Draw to make these amounts of money.

a. 63¢	b. 38¢	c. 69¢

3. You buy an item. How much money will you have left?

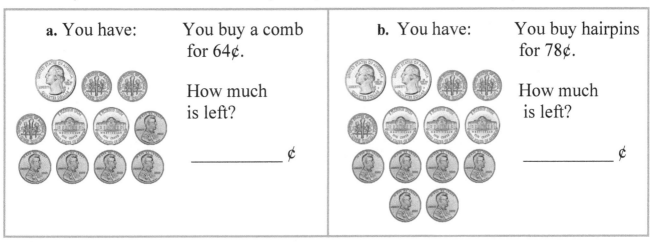

a. You have: You buy a comb for 64¢.

How much is left?

_____ ¢

b. You have: You buy hairpins for 78¢.

How much is left?

_____ ¢

End-of-the-Year Test - Grade 1

This test is quite long, so I do not recommend that you have your child/student do it in one sitting. Break it into parts and administer them either on consecutive days, or perhaps on morning/evening/morning. Use your judgment.

This is to be used as a diagnostic test. Thus, you may even skip those areas and concepts that you already know for sure your student has mastered.

The test does not cover every single concept that is covered in *Math Mammoth Grade 1*, but all of the major concepts and ideas are tested here. This test is evaluating the child's ability in the following content areas:

- basic addition and subtraction facts within 0-10
- two-digit numbers
- adding and subtracting two-digit numbers
- basic word problems
- clock to the nearest half hour
- measuring and geometry (shapes)
- counting coins

Note 1: If the child cannot read, the teacher can read the questions.

Note 2: Problems #1 and #2 are done <u>orally and timed</u>. Let the student see the problems. Read each problem aloud, and wait a maximum of 5 seconds for an answer. Mark the problem as right or wrong according to the student's (oral) answer. Mark it wrong if there is no answer. Then you can move on to the next problem.

You do not have to mention to the student that the problems are timed or that he/she will have 5 seconds per answer, because the idea here is not to create extra pressure by the fact it is timed, but simply to check if the student has the facts memorized (quick recall). You can say for example (vary as needed):

"I will ask you some addition and subtraction questions. Try to answer them as quickly as possible. In each question, I will only wait a little while for you to answer, and if you don't say anything, I will move on to the next problem. So just try your best to answer the questions as quickly as you can."

In order to continue with the Math Mammoth Grade 2, I recommend that the child gain a minimum score of 80% on this test, and that the teacher or parent review with him any content areas that are found weak. Children scoring between 70 and 80% may also continue with grade 2, depending on the types of errors (careless errors or not remembering something, vs. lack of understanding). Again, use your judgment.

Instructions to the student:

Answer each question in the space provided.

Instructions to the teacher:

My suggestion for grading is below. The total is 104 points. A score of 83 points is 80%. A score of 73 points is 70%.

Question	Max. points	Student score
Basic Addition and Subtraction Facts within 0-10		
1	8 points	
2	8 points	
3	4 points	
4	8 points	
	subtotal	/ 28
Place Value and Two-Digit Numbers		
5	6 points	
6	4 points	
7	3 points	
	subtotal	/ 13
Adding and Subtracting Two-Digit Numbers		
8	6 points	
9	6 points	
10	4 points	
11	3 points	
	subtotal	/ 19

Question	Max. points	Student score
Basic Word Problems		
12	2 points	
13	2 points	
14	2 points	
15	2 points	
16	2 points	
17	6 points	
18	6 points	
	subtotal	/ 22
Clock		
19	6 points	
20	4 points	
	subtotal	/ 10
Geometry and Measuring		
21	2 points	
22	5 points	
	subtotal	/ 7
Money		
23	3 points	
24	2 points	
	subtotal	/ 5
TOTAL		**/ 104**

End-of-the-Year Test - Grade 1

Basic Addition and Subtraction Facts within 0-10

In problems 1 and 2, your teacher will read you the addition and subtraction questions. Try to answer them as quickly as possible. In each question, he/she will only wait a little while for you to answer, and if you don't say anything, your teacher will move on to the next problem. So, just try your best to answer the questions as quickly as you can.

1. Add.

a.	b.	c.	d.
2 + 3 = _____	7 + 3 = _____	6 + 2 = _____	5 + 5 = _____
4 + 4 = _____	5 + 4 = _____	4 + 6 = _____	2 + 4 = _____
1 + 6 = _____	3 + 6 = _____	2 + 5 = _____	9 + 1 = _____
2 + 7 = _____	1 + 7 = _____	6 + 2 = _____	5 + 3 = _____

2. Subtract.

a.	b.	c.	d.
8 – 3 = _____	5 – 3 = _____	7 – 3 = _____	10 – 3 = _____
6 – 4 = _____	7 – 4 = _____	9 – 4 = _____	5 – 4 = _____
10 – 6 = _____	9 – 6 = _____	4 – 3 = _____	8 – 6 = _____
8 – 7 = _____	6 – 3 = _____	10 – 7 = _____	9 – 7 = _____

3. Write a fact family to match the picture.

_____ + _____ = _____   _____ + _____ = _____

_____ – _____ = _____   _____ – _____ = _____

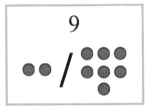

40

4. Find the missing numbers.

a. $2 + \underline{\hspace{1cm}} = 7$	b. $1 + \underline{\hspace{1cm}} = 8$	c. $4 + \underline{\hspace{1cm}} = 6$	d. $\underline{\hspace{1cm}} + 3 = 8$
$3 + \underline{\hspace{1cm}} = 8$	$2 + \underline{\hspace{1cm}} = 10$	$\underline{\hspace{1cm}} + 3 = 9$	$\underline{\hspace{1cm}} + 6 = 10$

Place Value and Two-Digit Numbers

5. Fill in the missing parts.

a. $20 + 7 = \underline{\hspace{1.5cm}}$	b. $6 + \underline{\hspace{1.5cm}} = 56$	c. $40 + \underline{\hspace{1.5cm}} = 40$
$5 + 60 = \underline{\hspace{1.5cm}}$	$30 + \underline{\hspace{1.5cm}} = 39$	$4 + \underline{\hspace{1.5cm}} = 94$

6. Put the numbers in order.

a. 16, 61, 26	b. 54, 14, 51
$\underline{\hspace{1.5cm}} < \underline{\hspace{1.5cm}} < \underline{\hspace{1.5cm}}$	$\underline{\hspace{1.5cm}} < \underline{\hspace{1.5cm}} < \underline{\hspace{1.5cm}}$

7. Compare the expressions and write $<$, $>$, or $=$.

    a. $40 + 8 \;\boxed{}\; 4 + 80$ **b.** $43 + 5 \;\boxed{}\; 50$ **c.** $3 + 33 \;\boxed{}\; 36$

Adding and Subtracting Two-Digit Numbers

8. Add.

a. $84 + 4 = \underline{\hspace{1.5cm}}$	b. $6 + 70 = \underline{\hspace{1.5cm}}$	c. $74 + 5 = \underline{\hspace{1.5cm}}$
$41 + 4 = \underline{\hspace{1.5cm}}$	$16 + 2 = \underline{\hspace{1.5cm}}$	$6 + 53 = \underline{\hspace{1.5cm}}$

9. Subtract.

a. $80 - 30 = \underline{\hspace{1.5cm}}$	b. $55 - 3 = \underline{\hspace{1.5cm}}$	c. $29 - 3 = \underline{\hspace{1.5cm}}$
$17 - 3 = \underline{\hspace{1.5cm}}$	$100 - 40 = \underline{\hspace{1.5cm}}$	$50 - 2 = \underline{\hspace{1.5cm}}$

10. Add and subtract.

a. $\begin{array}{r} 1\ 4 \\ +\ 3\ 5 \\ \hline \end{array}$
b. $\begin{array}{r} 5\ 9 \\ -\ 3\ 4 \\ \hline \end{array}$
c. $\begin{array}{r} 4\ 0 \\ +\ 5\ 6 \\ \hline \end{array}$
d. $\begin{array}{r} 9\ 6 \\ -\ 6\ 0 \\ \hline \end{array}$

11. Add. The images can help you.

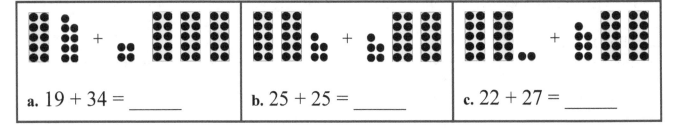

a. 19 + 34 = _____

b. 25 + 25 = _____

c. 22 + 27 = _____

Basic Word Problems

12. Write a subtraction sentence that matches with the addition 6 + 8 = 14.

_____ − _____ = _____

13. How many more is 70 than 50? _____ more

14. Henry owns four more cars than Mark, and Mark owns six cars.
    Draw Mark's cars and Henry's cars.

15. Ten kids are playing in the yard. There are 6 boys. How many girls are there?

16. Andy had 20 dollars. He bought a book for 10 dollars and another for 5 dollars.
    How much money does he have left?

17. A parking lot has spaces for 30 cars. There are cars in 22 of those spaces.

    a. How many spaces are empty?

    b. Two more cars arrive and park. Now how many cars are parked in the lot?

    c. How many empty spaces are there now?

18. Isabelle had 70 marbles and her sister had 55. Isabelle gave 10 marbles to her sister.

    a. Now how many marbles does Isabelle have?

    b. How many marbles does her sister have now?

    c. Who has more? How many more?

Clock

19. Write the time in two ways: using *o'clock* or *half past*, and with numbers.

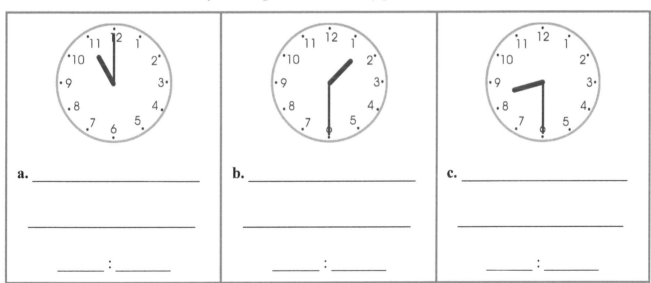

    a. _____

    _____

    _____ : _____

    b. _____

    _____

    _____ : _____

    c. _____

    _____

    _____ : _____

20. Write the time for a half-hour and an hour later from the given time. Use numbers.

Now it is:	**a.** 5:30	**b.** 12:00
a half-hour later, it is:		
an hour later, it is:		

Geometry and Measuring

21. Draw a line that is:

a. 3 inches

b. 9 centimeters

22. **a.** Join these dots carefully with a ruler so that you get a shape.

A • • B

D • • C

b. What is this shape called? ____________________________________

c. Measure the sides of your shape in centimeters.

Side AB: ____________ cm Side BC: ____________ cm

d. Draw a straight line from dot A to dot C. The line divides your shape into two new shapes.

What are the new shapes called? ________________________________

Money

23. How much money? Write the amount in cents.

a. _____ ¢

b. _____ ¢

c. _____ ¢

24. Solve.

You have:

You bought an apple for 35¢.

How much money do you have left? _____ ¢

Using the Cumulative Reviews

The cumulative reviews practice topics in various chapters of the Math Mammoth complete curriculum, up to the chapter named in the review. For example, a cumulative review for chapters 1-6 may include problems matching chapters 1, 2, 3, 4, 5, and 6. The cumulative review lesson for chapters 1-6 can be used any time after the student has studied the curriculum through chapter 6.

These lessons provide additional practice and review. The teacher should decide when and if they are used. The student doesn't have to complete all the cumulative reviews. I recommend using at least three of these reviews during the school year. The teacher can also use the reviews as diagnostic tests to find out what topics the student has trouble with.

Math Mammoth complete curriculum also includes an easy worksheet maker, which is the perfect tool to make more problems for children who need more practice. The worksheet maker covers most topics in the curriculum, excluding word problems. Most people find it to be a very helpful addition to the curriculum.

The download version of the curriculum comes with the worksheet maker, and you can also access the worksheet maker online at

https://www.mathmammoth.com/private/Make_extra_worksheets_grade1.htm

Cumulative Review, Grade 1, Chapters 1-2

1. Draw arrows to show the addition and the subtractions.

a. 6 + 2 = _____

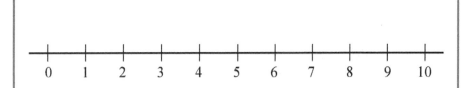

b. 9 − 4 = _____

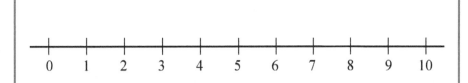

c. 7 − 3 = _____

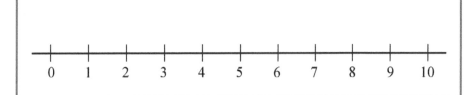

2. Make an addition sentence and a subtraction sentence from the same picture.

a. ●●●●○○○○○○

_____ + _____ = _____

_____ − _____ = _____

b. ☐☐☐☐☐▧▧▧

_____ + _____ = _____

_____ − _____ = _____

3. Add. Remember, you can add in any order.

a.	b.	c.	d.	e.
2	5	2	6	4
1	1	1	1	0
+ 4	+ 4	+ 1	+ 2	+ 3

4. Fill in the missing numbers.

| a. 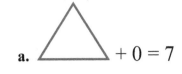 $+ 0 = 7$ | b. $\triangle - 2 = 4$ | c. $8 - \triangle = 1$ |

5. Compare. Write $<$, $>$, or $=$.

a. $2 + 3 \ \square \ 5$

b. $6 + 1 \ \square \ 8$

c. $7 - 1 \ \square \ 9$

d. $4 - 4 \ \square \ 0$

e. $2 \ \square \ 4 - 2$

f. $9 \ \square \ 9 - 1$

6. Find the missing numbers.

a.	b.	c.	d.
$6 + 4 = \underline{\hphantom{00}}$	$2 + 7 = \underline{\hphantom{00}}$	$3 + \underline{\hphantom{00}} = 10$	$2 + \underline{\hphantom{00}} = 9$
$4 + 4 = \underline{\hphantom{00}}$	$5 + 3 = \underline{\hphantom{00}}$	$3 + \underline{\hphantom{00}} = 8$	$2 + \underline{\hphantom{00}} = 7$

7. Draw the missing marbles to match the addition sentence.

a. $6 + 1 + \underline{\hphantom{00}} = 10$

b. $1 + 4 + \underline{\hphantom{00}} = 8$

8. Write the fact families.

a. Numbers: 9, 5, 4

$\underline{\hphantom{00}} + \underline{\hphantom{00}} = \underline{\hphantom{00}}$

$\underline{\hphantom{00}} + \underline{\hphantom{00}} = \underline{\hphantom{00}}$

$\underline{\hphantom{00}} - \underline{\hphantom{00}} = \underline{\hphantom{00}}$

$\underline{\hphantom{00}} - \underline{\hphantom{00}} = \underline{\hphantom{00}}$

b. Numbers: 10, 2, 8

$\underline{\hphantom{00}} + \underline{\hphantom{00}} = \underline{\hphantom{00}}$

$\underline{\hphantom{00}} + \underline{\hphantom{00}} = \underline{\hphantom{00}}$

$\underline{\hphantom{00}} - \underline{\hphantom{00}} = \underline{\hphantom{00}}$

$\underline{\hphantom{00}} - \underline{\hphantom{00}} = \underline{\hphantom{00}}$

9. Draw marbles for the child that has none.

	Jane	◉◉◉◉◉◉◉◉	Luis
◉◉◉◉◉	Greg		Henry

a. Jane has 2 more than Greg. **b.** Luis has 4 more than Henry.

	Jill		Jim
◉◉◉◉◉◉◉	Bill	◉◉◉◉	Ann

c. Jill has 2 fewer than Bill. **d.** Ann has 3 fewer than Jim.

10. Solve.

a. Three children were playing. Then, five more children came to play. Then, one child left. Now how many children are playing?

b. Judy has 3 marbles and Annie has 7.
How many marbles do the girls have together?

How many more does Annie have than Judy?

c. Kyle has 10 toy trucks. Some are blue and seven are black.
How many are blue?

d. Leah has 6 dollars. She wants to buy a book for $9.
How much more money does she need?

e. Matt has 10 socks and he can't find any of them! Then, he found three socks under the bed and five in the closet. How many socks did Matt find?

How many are still missing?

Cumulative Review, Grade 1, Chapters 1-3

1. Break the numbers into tens and ones.

a. 22 = _____ + _____	**b.** 64 = _____ + _____	**c.** 95 = _____ + _____

2. Compare. Write < , > , or = .

a. 2 + 3 ☐ 5 + 1	**c.** 8 − 2 ☐ 4	**e.** 6 ☐ 4 + 2
b. 6 + 4 ☐ 8 + 2	**d.** 7 − 4 ☐ 5	**f.** 8 ☐ 9 − 1

3. Write the fact families.

a. 3 + 7 = _____ _____ + _____ = _____ _____ − _____ = _____ _____ − _____ = _____	**b.** 6 + _____ = 9 _____ + _____ = _____ _____ − _____ = _____ _____ − _____ = _____

4. Skip-count by tens.

    a. 4, 14, _____, _____, _____, _____, _____, _____

    b. _____, _____, _____, 68, 78, _____, _____, _____

5. **a.** Skip-count by fives starting at 45, and color all those numbers yellow.

    b. Skip-count by twos starting at 42, and color those numbers blue.

41	42	43	44	45	46	47	48	49	50
51	52	53	54	55	56	57	58	59	60
61	62	63	64	65	66	67	68	69	70

    Which numbers end up green?

6. Name and write the numbers.

   a. 1 ten 1 one _____

   b. 1 ten 7 ones _____

   c. 1 ten 5 ones _____

   d. 1 ten 3 ones _____

7. Find the difference between the numbers. "Travel" on the number line!

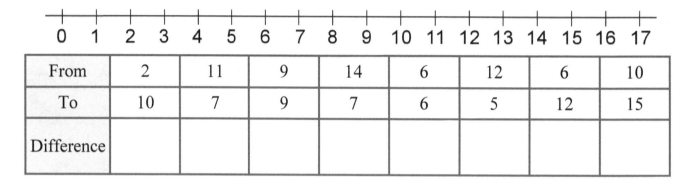

From	2	11	9	14	6	12	6	10
To	10	7	9	7	6	5	12	15
Difference								

8. Write < or > or = .

a. 82 ☐ 29	**c.** 70 + 4 ☐ 7 + 40	**e.** 60 + 7 ☐ 70 + 5
b. 75 ☐ 67	**d.** 20 + 8 ☐ 2 + 80	**f.** 2 + 90 ☐ 9 + 50

9. Solve.

   a. Some children needed 10 players for a game. They already had 2 boys and 2 girls. How many more children do they need for their game?

   b. A herd has 10 brown horses, 20 white horses, and 10 speckled ones. How many horses are there in the herd?

     How many more white horses are there than brown ones?

Cumulative Review, Grade 1, Chapters 1-4

1. Pick a number so the comparison is true.

3 4 5	4 5 6	3 4 5
2 + ____ < 6	1 + ____ > 6	4 + ____ < 8

2. Add.

   a. $0 + 4 + 2 =$ _____   **b.** $7 + 1 + 1 =$ _____   **c.** $2 + 5 + 3 =$ _____

3. Fill in the numbers and name them.

a. _____	**b.** _____	**c.** _____
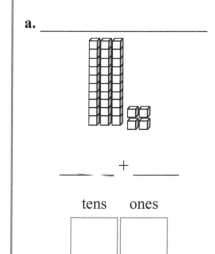	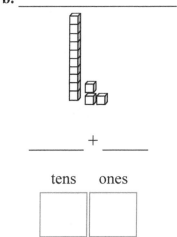	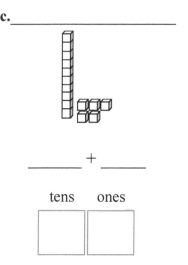
_____ + _____	_____ + _____	_____ + _____
tens ones	tens ones	tens ones

4. The numbers are broken into tens and ones. Fill in the missing parts.

a. 40 + ____ = 48	**b.** ____ + ____ = 62	**c.** 50 + 5 = ____

5. What number is...

a.	**b.**	**c.**
one more than 16 _____	two more than 11 _____	ten more than 12 _____
one less than 29 _____	two less than 67 _____	ten less than 30 _____
one less than 40 _____	two more than 59 _____	ten more than 87 _____

6. Count. You can also do this orally with your teacher.

96, 97, _____, _____, _____, _____, _____,

_____, _____, _____, _____, _____, _____

7. Draw tally marks for these numbers.

a. 9	**b.** 11
c. 27	**d.** 32

8. Some children counted how many stuffed animals they had.

    a. How many does Alice have?

    b. How many does Aaron have?

    c. How many does Maria have?

    d. Alice gave 10 stuffed
       animals to Aaron. Now
       how many does Alice
       have?

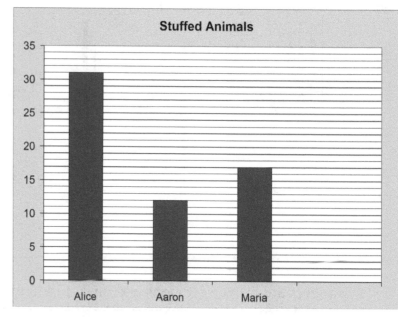

    And Aaron?

    In the empty space you can draw a bar on the graph for how many stuffed animals you have.

9. Find the "mystery numbers"! You will need to think logically.

a. This number has three more tens than 50 has, and the same amount of ones as 13.	**b.** This number has seven less ones than 29, and six more tens than 17.

56

Cumulative Review, Grade 1, Chapters 1 - 5

1. Add.

| **a.** $1 + 4 =$ _____ | **b.** $5 + 2 =$ _____ | **c.** $3 + 6 =$ _____ | **d.** $7 + 3 =$ _____ |

2. Subtract.

a. $5 - 2 =$ _____   **b.** $9 - 4 =$ _____   **c.** $7 - 3 =$ _____   **d.** $10 - 8 =$ _____

e. $6 - 2 =$ _____   **f.** $10 - 7 =$ _____   **g.** $7 - 7 =$ _____   **h.** $9 - 5 =$ _____

3. Write the names of the numbers with whole tens.

two tens   _____

three tens   _____

eight tens   _____

five tens   _____

4. Put the numbers in order from the smallest to the largest.

a. 58, 17, 36	**b.** 23, 63, 36	**c.** 48, 84, 44
_____ < _____ < _____	_____ < _____ < _____	_____ < _____ < _____

5. Solve the missing numbers.

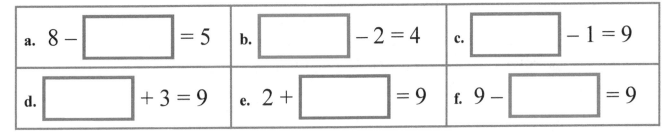

a. $8 - \boxed{} = 5$	**b.** $\boxed{} - 2 = 4$	**c.** $\boxed{} - 1 = 9$
d. $\boxed{} + 3 = 9$	**e.** $2 + \boxed{} = 9$	**f.** $9 - \boxed{} = 9$

6. Draw the hour hand on the clocks. Then write the time that the clock will show a half-hour later.

	a. two o'clock	b. ten o'clock	c. half-past six	d. half-past eight
1/2 hour later →	_____ _____	_____ _____	_____ _____	_____ _____

7. Fill in either AM or PM.

a. You woke up. It was 7 _____.	b. Jon plays in the afternoon at 3 _____.
c. Joe is asleep. It is dark. It is 1 _____.	d. It is time for lunch. It is 1 _____.

8. Compare. Write < , > , or = .

a. 62 ☐ 3 + 60	c. 10 − 2 ☐ 7	e. 7 ☐ 9 − 2
b. 54 ☐ 42 + 10	d. 6 − 5 ☐ 0	f. 45 ☐ 65 − 10

9. Solve.

a. A baby put some of his ten crayons into a bucket. Then he had 4 crayons on the floor. So, how many did he put into the bucket?

b. Theresa has $10. She gets another $4 from her mom.
Now how much does she have?

How much more does she need to buy a $20 shirt?

Cumulative Review, Grade 1, Chapters 1 - 6

1. Find the missing numbers.

| **a.** $7 + \underline{\hspace{1cm}} = 7$ | **b.** $\underline{\hspace{1cm}} + 6 = 9$ | **c.** $4 + \underline{\hspace{1cm}} = 9$ | **d.** $8 + \underline{\hspace{1cm}} = 10$ |

2. Cross out the problem if you can't take away that many.

$4 - 6$ $2 - 0$ $6 - 9$ $8 - 4$

3. Write these numbers with words. You can also do this orally with your teacher.

   a. 2 tens 9 ones = _____

   b. 9 tens 1 one = _____

   c. 1 ten 5 ones = _____

   d. 5 tens 7 ones = _____

4. Write the time for a half-hour later. Use numbers.

Now it is:	**a.** 1:00	**b.** 11:30	**c.** 9:00	**d.** 6:30	**e.** 4:00
a half-hour later, it is:	___ : ____	___ : ____	___ : ____	___ : ____	___ : ____

5. Draw different shapes that you have learned, and label them.

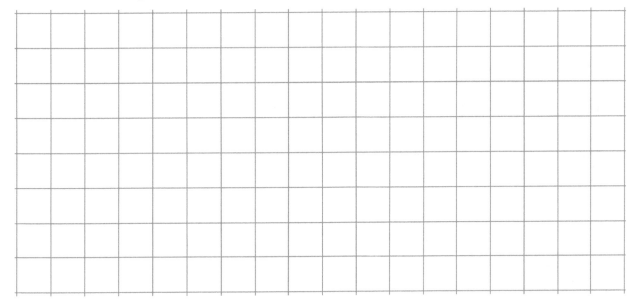

6. Color the circles yellow,
the triangles blue,
the squares pink,
the rectangles green
and the rest of the shapes red.

7. Solve.

9 − _____ = 4

3 + _____ = 4

10 − _____ = 4

2 + _____ = 4

3 + _____ = 7

8 − _____ = 7

0 + _____ = 7

9 − _____ = 7

10 − _____ = 8

3 + _____ = 8

9 − _____ = 8

2 + _____ = 8

9 − _____ = 6

3 + _____ = 6

10 − _____ = 6

2 + _____ = 6

3 + _____ = 5

10 − _____ = 5

0 + _____ = 5

9 − _____ = 5

10 − _____ = 3

3 + _____ = 3

7 − _____ = 3

2 + _____ = 3

8. Write < or > between the numbers to compare them.

a. 30 < 38   b. 87 ☐ 85   c. 69 ☐ 96   d. 58 ☐ 56

e. 60 ☐ 48   f. 43 ☐ 95   g. 49 ☐ 94   h. 22 ☐ 32

60

Cumulative Review, Grade 1, Chapters 1 - 7

1. Find the missing numbers.

a.	b.	c.	d.
2 + _____ = 4	2 + _____ = 7	6 − _____ = 6	3 − 1 = _____
5 + _____ = 9	0 + _____ = 5	8 − _____ = 4	10 − 3 = _____

2. Find the "mystery numbers"! You will need to think logically.

a. This number has one more ten than 20 has, and the same amount of ones as 63.	**b.** This number has two less ones than 88, and six more tens than 24.

3. Count by tens.

16, 26, _____, _____, _____, _____, _____, _____

4. Find how much the two items cost together.

a. a rabbit, $12, and a parrot, $65 Together they cost $ _____ .	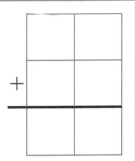	b. a bicycle, $76, and a flashlight, $23 Together they cost $ _____ .	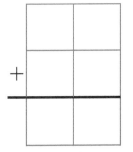

5. Draw the hands to show the time.

	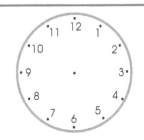	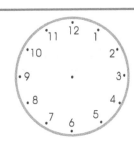	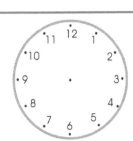
a. 5 o'clock	**b.** half past seven	**c.** 11 o'clock	**d.** half past two

6. Fill in the blanks.

    a. _____ has four sides the same length.

    b. _____ has three sides and three vertices.

7. Add using the nine trick and the eight trick.

a. $9 + 8 =$ _____ $8 + 8 =$ _____	**b.** $9 + 3 =$ _____ $8 + 4 =$ _____	**c.** $9 + 5 =$ _____ $7 + 8 =$ _____

8. Subtract and add whole tens.

a. $25 + 10 =$ _____ $60 + 20 =$ _____	**b.** $90 - 30 =$ _____ $100 - 70 =$ _____	**c.** $92 - 10 =$ _____ $64 - 10 =$ _____

9. Divide these shapes by drawing straight lines from dot to dot. Then color.

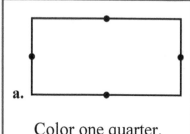 **a.** Color one quarter.	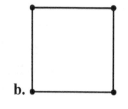 **b.** Color two quarters.	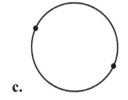 **c.** Color two halves.	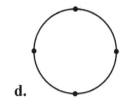 **d.** Color three fourths.

10. Which is more, two quarters or one half?
   Coloring the parts in the pictures may help.

 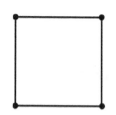

11. Name the basic shape. Is it a cylinder, a cube, a box, or a ball?

 **a.**	 **b.**	 **c.**	 **d.**

Cumulative Review, Grade 1, Chapters 1 - 8

1. Find the missing numbers.

a.	b.	c.	d.
_____ – 1 = 9	8 + 9 = _____	25 – 10 = _____	52 + 7 = _____
_____ – 2 = 6	7 + 8 = _____	38 – 10 = _____	35 + 3 = _____
_____ – 3 = 4	5 + 6 = _____	100 – 10 = _____	26 + 2 = _____

2. Draw the hour hand.

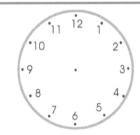 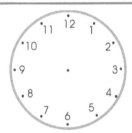 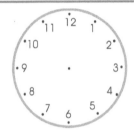

a. half past one	**b.** half past five	**c.** half past ten

3. Trace and cut the shapes to do the exercises, or
   if you can, imagine putting the shapes together,
   and sketch (draw) the answer shapes.

   a. Put together the #5 triangles so that you get
   a rectangle.

   b. Now place them together so that you get a
   different four-sided shape (not a rectangle).

   c. Use some of the #1 (yellow) triangles to form
   the purple shape (#6)

   d. Now use a #1 triangle and the purple shape.
   What shapes can you make with these two?

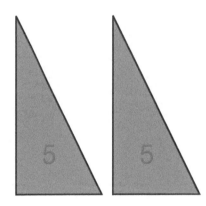

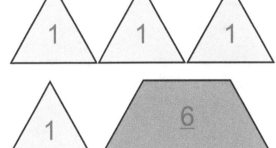

4. First subtract to 10. Then subtract the rest.

| a. 15 – 7
 / \
 15 – ____ – ____

 = ____ | b. 14 – 9
 / \
 14 – ____ – ____

 = ____ | c. 16 – 8
 / \
 16 – ____ – ____

 = ____ |

5. Add and subtract.

a. $\begin{array}{r} 2\ 5 \\ +\ 3\ 0 \\ \hline \end{array}$ b. $\begin{array}{r} 5\ 1 \\ +\ 3\ 4 \\ \hline \end{array}$ c $\begin{array}{r} 7\ 8 \\ -\ 1\ 5 \\ \hline \end{array}$ d. $\begin{array}{r} 8\ 6 \\ -\ 4\ 4 \\ \hline \end{array}$

6. Draw circles to make these amounts of money. Put **P** on pennies, **D** on dimes, **N** on nickels and **Q** on quarters. Use the least number of coins possible.

| a. 32¢ | b. 15¢ | c. 26¢ |
| d. 80¢ | e. 66¢ | f. 43¢ |

7. You bought an item. How much money do you have left?

| a. You have:  | You bought a bar of soap for 59¢.

 How much is left?

 _____ ¢ | b. You have:  | You bought a toy for 86¢.

 How much is left?

 _____ ¢ |